CW00549014

AIRLINE

BOEING 727

Adrian Balch

LUFTHANSA

Europa Jet 727

Acknowledgements

I would like to thank the many photographers, whose names are credited under each picture. Please accept my apologies where the photographer is unknown.

Special thanks go to Tom Cole of The Boeing Commercial Airplane Co. and to Christel Zwartjes of Lufthansa's Public Relations Dept., also to United Airlines for photographs which they have supplied.

Title page photograph shows 727-30, D-ABIN (c/n 18369) in this original livery, as delivered in March 1965.
(Photo: Deutsche Lufthansa).

Copyright © Adrian Balch, 1992

First published in the UK in 1992 by Airlife Publishing Ltd.

British Library Cataloguing in Publication Data.
A catalogue record for this book
is available from the British Library.

ISBN 1 85310 341 1

All rights reserved. No part of this book may be reproduced or transmitted in any form or by any means, electronic or mechanical including photocopying, recording or by any information storage and retrieval system, without permission from the Publisher in writing.

Printed in Singapore by Kyodo Printing Co (S'pore) Pte Ltd

Airlife Publishing Ltd.

101 Longden Road, Shrewsbury, SY3 9EB, England.

Introduction

The Boeing 727 was the first tri-jet introduced to commercial use and became the best selling civil airliner of its time. Following discussions with many US airlines, the type was officially unveiled on 5 December 1960, with the first orders coming from Eastern and United Air Lines. They each ordered forty of the 727-100 variant. Boeing estimated they would sell a maximum of 250 of the type, but ended up selling 1832. Production ceased in August 1984, with 584 sold to overseas customers and the rest to US customers. In 1965, 193 were ordered, but sales dropped in 1981 to thirty-eight and finally just one in 1983. Fuel economy on medium-range routes was what helped bring production to an end, as the airlines needed something more economical than a tri-jet, and this came in the form of the Boeing 757. However, by mid-1983 Boeing 727s were carrying 43 million passengers per month. The first 727-100, a production model, made its first flight from Renton Field on 9 February 1963, which was followed by FAA certification on Christmas Eve of that year. The type started scheduled services with Eastern Air Lines on 1 February 1964, followed five days later by United Air Lines. Several variants of the 727 were built, with the -100 series being the standard passenger transport for up to 131 passengers, while -100C was a convertible cargo passenger version, announced in July 1964. The 727-100QC was the same as the latter, except that by using palletised passenger seats, galleys and advanced cargo loading techniques, the aircraft could easily be converted from an all-cargo to an all-passenger configuration and vice-versa in half an hour. In 1965, a 'stretched' version was announced, with basic accommodation for 178 passengers. The 727-200 had a fuselage extension of 3.05m (10ft), as well as more powerful engines. This variant first flew in May 1967 and received its FAA certification in September of that year. First delivery was to Northeast airlines in June 1972 and by September of that year orders had reached 1000. Pan Am took delivery of the 1800th example with US Air receiving the final 727 off the production line on 6 April 1983. In 1970, the Advanced 727-200 was announced, with extended range, increased gross weight and a new 'wide-body' look to the interior. Many airlines have displaced their 727-10s with 727-200s and some operate both, giving testimony to the design.

The Boeing 727 owes much of its success to the design of the Sud-Aviation Caravelle, which was the first jet airliner with its engines at the rear of the fuselage and an integral folding staircase under the rear of the fuselage. It is these features that contribute to the type's popularity with travellers. Also, it will be noted that the cockpit section has much in common with the pioneering Boeing 707, so the total package had a good pedigree, and was almost assured of the success it has had with the airlines.

TABLE OF COMPARISONS		
	-100	-200
Max. accommodation:	131	178
Wing span:	32.92 m (108 ft 0 in)	32.92 m (108 ft 0 in)
Length:	40.59 m (133 ft 2 in)	46.69 m (153 ft 2in)
Height:	10.36 m (34 ft 0 in)	10.36 m (34 ft 0 in)
Max. t/o weight:	64,410 kg (142,000 lb)	76,655 kg (169,000 lb)
Max. cruise speed:	974 km/h (605 mph)	953 km/h (592 mph)
Maximum range:	4330 km (2,690 miles)	3,700 km (2,300 miles)
Service ceiling:	11,400 m (37,400 ft)	10,730 m (35,200 ft)

ARIANA-AFGHAN AIRLINES AFGHANISTAN

Ariana, formed in 1955 by the Government of Afghanistan (51 per cent) and Pan American (49 per cent) was operating DC-6ABs and Convair 440s until the first 727 was introduced in April 1968 on routes from Kabul to London, Frankfurt, Beirut, Tehran, Tashkent, Moscow and New Delhi. This aircraft was destroyed in a landing accident at Gatwick in January 1969, but a replacement was ordered almost immediately. A World Airways 727 was leased pending delivery of the new aircraft. An additional aircraft was purchased in June 1971 and two 727-100Cs are still in service with the airline.

The current colour scheme is a two-tone blue cheat line separated by a thin white line through the windows. The tail has a stylisted blue 'A' insignia encompassing the company's flying bird insignia. Ariana's original Boeing 727-113, YA-FAR(c/n 19690) lasted less than a year in service, as this ill-fated machine was delivered to

Ariana on 25 March 1968, but crashed on 5 January 1969 1½ miles short of Gatwick's runway in fog.

Forty-eight passengers were killed, plus two on the ground, there were fifteen survivors – the aircraft had flown 1,733 hours.

Ariana's original colour scheme on its 727s comprised a medium-blue cheatline, which started at a point on the nose and became thicker, until it swept up the fin, dividing it in two diagonally. 'ARIANA Afghan Airlines' appeared on the roof in English and Afghan script in the same blue and the fin sported the airline's white bird insignia.

The photograph shows 727-113C, YA-FAU (c/n 20343), which was Ariana's second 727, seen on a pre-delivery test flight in the airline's old colours in January 1970. (Boeing)

AIR ALGERIE

ALGERIA

Air Algerie is Algeria's designated carrier and scheduled passenger and cargo services are operated to thirty-four points in North and West Africa, France, Belgium, Spain, Italy, Greece, Austria, the UK, Germany, Switzerland, Egypt, Yugoslavia, Bulgaria, Russia, Libya, Romania, Czechoslovakia and the Middle East. An extensive domestic network is also operated, together with charter, air taxi and agricultural flights.

Air Algerie was formed in June 1953 by the merger of the original Air Algerie (founded in 1949) and Compagnie Air Transport. It became wholly Government owned in 1972.

The airline currently operates ten 727-200 series, which are the largest aircraft in the fleet and which replaced Caravelles on routes to France, Spain, Italy, Germany, Switzerland and Russia.

The current colour scheme is based on an overall white fuselage, which has red and green stripes running below the windows. Titles in English and Arabic appear in red on the roof, and the airline's arrow insignia appears on the fin in red.

The photograph shows 727-2D6,7T-VEW (c/n 22375) in Air Algerie's old livery during a pre-delivery test flight over Seattle, Washington in March 1981. (Boeing)

ANSETT AIRLINES OF AUSTRALIA   AUSTRALIA

Ansett is one of Australia's two major domestic carriers and is a subsidiary of Ansett Transport Industries (ATI). The company operates an extensive network of scheduled passenger services covering all the states of the Commonwealth of Australia.

The airline was formed in 1936 by R.M.Ansett as Ansett Airways. The 727 went into service in Australia on 7 November 1964 on the Sydney – Melbourne and Sydney – Adelaide – Perth runs. The first 727-200A went into service on 1 March 1973. Since Ansett first introduced the 727, they have changed their colour scheme four times. The current scheme is white overall with coloured shooting stars representing a stylised Australian flag on a dark blue fin.

The first photograph illustrates Ansett's old livery on 727-277, VH-RML (c/n 21480) over Washington, prior to delivery in June 1978 (Boeing). The second photograph illustrates the next scheme on 727-277,VH-ANA (c/n 22641) prior to delivery in June 1981. (Boeing)

AUSTRALIAN AIRLINES

AUSTRALIA

Australian Airlines is one of Australia's two major domestic carriers and is fully owned by the Australian Government. It operates an extensive network of passenger and cargo services, linking all state capital cities and many provincial towns. The company was formed in 1945 and began operations in September 1946 as Trans Australia Airlines. The name Australian Airlines was adopted in 1986.

Both Ansett and Australian Airlines (ex TAA) are required to purchase the same equipment and introduce them simultaneously.

Both airlines have consequently operated both versions of the Boeing 727, the first of which were ordered on 8 February 1963. The current colour scheme is based on an overall white fuselage with stripes of green and orange running along the lower half. The underside of the fuselage is dark blue, as is the kangaroo symbol on the fin. Australian Airlines operates ten Boeing 727-276 Advanced models, of which VH-TBN (c/n 21479) is seen here taxying at Sydney on 10 October 1988 (Werner Gysin-Aegerter).

AEROLINEAS ARGENTINAS ARGENTINA

Aerolineas Argentinas, the country's flag carrier, operates an extensive network of scheduled passenger and cargo routes to points in North, South and Central America, New Zealand, the Far East, South Africa and Europe. The company was founded as a state corporation in May 1949 by the Ministry of Transport to take over operation of Fama, Alfa, Aeroposta and Zonda. These companies ceased operations in December 1949 and were merged to form Aerolineas Argentinas. The airline's first 727 was leased from Hughes Airwest in December 1977, but they currently own and operate eight 727-287s. Colour scheme includes a two-tone blue cheat line which starts at a point on the nose and gradually gets thicker towards the rear, sweeping up the fin to encompass a 'double-A' insignia.

The photograph shows B727-287, LV-MI0 (c/n 21690) seen on a pre-delivery test flight in April 1979. (Boeing)

BELGIUM

SABENA-BELGIAN WORLD AIRLINES

Sabena was founded in 1923 as a successor to SNETA and developed a European network, followed by routes to and within the Belgian Congo (now Zaire). The airline currently operates an extensive network of scheduled passenger and cargo services to points in Europe, the Middle and Far East, twenty-six destinations in Africa and to Montreal, New York, Atlanta, Chicago, Boston and Toronto.

Sabena ordered five 727-29s for African and Middle East routes, as well as some European routes. The first 727 service (to London) was operated on 15 June 1967, and regular services from Brussels to both London and Milan began on 26 June 1967. This was soon followed by services to the Middle East and East Africa. The 727s replaced Caravelles and two were, in turn, replaced by 737s in 1974. In January 1976 the last of Sabena's 727s were withdrawn from use, when two were sold to the Belgian Air Force.

The second, and last, colour scheme on Sabena's 727s is shown here on 727-29C, 00-STD (c/n 19403) at Copenhagen-Kastrup in September 1974. A simple blue cheat line and blue fin upon which is mounted Sabena's stylised 'S' motif. 00-STD was delivered on 22 July 1967 and became CB-02 with the Belgian Air Force in January 1976. (Author's collection)

LLOYD AEREO BOLIVIANO   BOLIVIA

Lloyd Aereo Boliviano operates a network of scheduled passenger and cargo services consisting of flights to twenty-one cities within Bolivia and fifteen international destinations, including Buenos Aires, Asuncion, Sao Paulo, Arica, Lima, Santiago, Manaus, Caracus, Cuzco, Panama, Salta, Rio de Janeiro, Belo Horizonte, Montevideo and Miami.

LAB was formed on 15 September 1925 and is now 99.9 per cent Government owned. In September 1968, LAB wanted to order three 727s to replace DC-6Bs on its major routes from Sao Paulo and Rio de Janeiro through Santa Cruz, Cochabamba, La Paz and on to Lima. Due to purchasing difficulties, one aircraft was ordered in August 1969 and delivered to Braniff (which at one time had a 13.4 per cent

interest in the airline), one in 1974 and one in 1975. Late in 1987, a 727-200A built for an undisclosed customer, who subsequently cancelled, was purchased. This is the first use of a 727-200 in the high altitude environment of the Andes. The colour scheme comprises a simple blue cheat line, red roof tiles and the company's logo on the fin below a small Bolivian flag.

The airline currently operates four 727s, the photograph depicting 727-78, CP-1223 (c/n 18795) taxying at La Paz in September 1986. This aircraft was originally delivered to British West Indian Airways in January 1965, sold to Braniff in April 1971, then bought by LAB in March 1975. (Martin Siegrist)

CRUZEIRO
BRAZIL

Cruzeiro (Servicos Aereos Cruzeiro do Sul) operates scheduled passenger and cargo services within Brazil and to Santa Cruz and La Paz (Bolivia), Montevideo (Uruguay) and Buenos Aires (Argentina). The company was founded officially on 1 December 1927, as Sindicato Condor. The airline became Servicio Aeros Condor in March 1941, when it became wholly Brazilian-owned with private capital. The present title was adopted in January 1943, following complete nationalisation. Cruzeiro absorbed Transportes Aereos Catarinese and Viacao Area Gaucha during 1967. Cruzeiro announced its orders for three 727s after four aircraft had been rolled out in their colours. In the end, only three were actually

ordered. Cruzeiro decided on the 727-100 rather than the Srs.200, as the latter, with the engines then available, could carry no greater payload out of Congonhas Airport, Sao Paulo (elevation 2,500 ft). Cruzeiro, 85 per cent owned by its employees, operates a fleet of seven 727s and eight 737s. Colour scheme features a two-tone blue cheat line, which curves down under the nose. The fin is predominately light blue on which Cruzeiro's diamond motif appears.

The photograph shows 727-1C3, PP-CJF (c/n 20419) at La Paz, Bolivia in September 1986. (Martin Siegrist)

TRANS BRAZIL

BRAZIL

Transbrasil Linhas Aeras, a Brazilian carrier, operates scheduled passenger and cargo services from Brasilia to points in south-east and north-east Brazil, and a cargo service to Miami.

Transbrasil was formed on 5 January 1955 as Sadia SA Transportes Aereas. Its aim was to carry fresh meat products from Concordia to the markets of Sao Paulo, using a DC-3.

Scheduled passenger services over the same routeing and from other points in the state of Santa Catarina were started in March 1956. The present name was adopted in 1972. Transbrasil initially purchased two 727Cs from Pan American. Transbrasil operated nine 727-100s, but all were sold in 1990 as more 737s were operated. This interesting photograph shows Transbrasil's old colour scheme at Sao Paulo in July 1978, with 727-27C, PT-TYU (c/n 19109) in orange and green. Two variations on this scheme are visible in two-tone green and maroon and biscuit. This scheme was probably influenced by Braniff's revolution in colour schemes (Author's collection)

VARIG

BRAZIL

Varig was founded in 1927, with the technical assistance of the German Condor syndicate, for the operation of local services. Varig subsequently absorbed several other airlines, including the domestic operator Aero Geral (1951), the REAL consortium (1961) and Panair do Brasil's equipment and international routes in 1965. Today, Varig operates an extensive network of routes throughout South and Central America, and to the USA, Europe, Africa and Japan. Domestic operators Cruzeiro and Rio-Sul are subsidiaries. Varig currently operates ten 727-100s on domestic routes. Colour scheme is a deep blue cheat line, divided by twin double white lines. Blue 'VARIG' titles and the Brazilian flag appear on the white roof. Varig's compass insignia appears on the white fin, with smaller 'Varig' titles beneath it.

The photograph shows 727-41, PP-VLD (c/n 20425) on a pre-delivery test flight over Washington in October 1970. (Boeing)

AIR CANADA

Air Canada was founded in 1937 as Trans-Canada Airlines. Operations began a year later between Vancouver and Seattle, with transcontinental passenger services commencing in April 1939. Today, the company operates an extensive route network that covers the USA, Bahamas, Cuba, Europe, India and Singapore. As well as international flights, a domestic network in association with various commuter carriers is operated. Cargo services are also maintained across the network. Air Canada's 727-233As were ordered on 31 December 1973 and the airline currently operates ten of the type on routes in Canada and the USA. Colour scheme is a broad red cheat line and red fin with the airline's Maple Leaf insignia in white. In recent years, black roof titles have been changed to red.

The photograph shows 727-233, C-GAAB (c/n 20933) seen over Washington on a pre-delivery test flight in September 1974. This aircraft was sold to Federal Express in December 1986. (Boeing)

CP AIR (CANADIAN PACIFIC AIR LINES) CANADA

Canadian Pacific Air Lines was Canada's second-largest airline, operating an extensive network of international and domestic routes until April, 1987, when it merged with Pacific Western Airlines to form Canadian Airlines International, Canadian Pacific was formed in 1942 when the railway company of this name acquired and amalgamated ten bush operators. Initially the airline operated services in north-west and western Canada but international services began in 1949. Eastern Provincial Airways was acquired in 1984 and in 1985 Nordair was acquired. When CP Air and Pacific Western Airlines merged, Canadian Airlines International came into being in January 1988.

The airline did not keep its 727s after the merger. The photograph shows CP Air 727-17, CF-CUR (c/n 20512) at Toronto on 2 June 1976 in the airline's smart natural metal, orange and red scheme. (John Kimberley)

CAYMAN AIRWAYS   CAYMAN ISLANDS

Cayman Airways operated scheduled passenger services linking George Town (Grand Cayman) with Kingston/Montego Bay, Miami, Houston Tampa and Atlanta. Local services are provided to Little Cayman and Cayman Brac. Passenger charter flights are also operated between the Cayman Islands and Atlanta, Baltimore, New York and Philadelphia. The airline was formed in July 1968 to replace Cayman Brac Airways, a subsidiary of the Costa Rican airline, LACSA. In December 1977, the airline became wholly Government owned and currently has a small fleet of two 727-200s and a Shorts 330. The colour scheme comprises a red/white/blue cheat line with titles in dark blue lower-case lettering. The fin features the airline's flying turtle motif.

The photograph shows 727-227 Advanced, N272AF (c/n 22004) on finals to Miami in January 1986. The registration suffix is retained from its previous owner – Air Florida. (Nigel Chalcraft)

AVIANCA   COLOMBIA

Avianca is one of the world's oldest airlines, tracing its history as far back as December 1919, with the formation of Sociedad Colombia-Alemana de Transportes Aereos. It was the first airline in the Americas, and was formed by a group of German settlers. Scheduled operations began two years later between Girardot and Barranquilla, using a fleet of Junkers F-13 floatplanes. Today the company provides an extensive network of operations that covers South America, the USA, Spain, France and West Germany. Avianca was the first South American airline to purchase the 727, which replaced Constellations and Super Constellations on some difficult domestic routes in Colombia, connecting Bogota (8,355 feet elevation), Barranquilla, Cali and Medellin (4,196 feet elevation), with surrounding mountains up to 11,000 feet) on 29 April 1966.

Avianca currently operates four 727-100s and seventeen 727-200s in their unusual and attractive colour scheme with red roof and white and black titles.

The photograph shows 727-2A1, HK-2151 (c/n 21343) photographed on a test flight near Seattle on 9 October 1978, just three days before delivery. (John Mounce).

ACES – AEROLINEAS CENTRALES DE COLOMBIA   COLOMBIA

Aerolineas Centrales de Colombia (ACES) operates domestic scheduled passenger services throughout Colombia with main hubs at Bogota, Medellin and Cali. Jet services link Bogota, Cali, Medellin, Barranquilla, Cartagena, Armenia and San Andres, and thirty-seven points receive twin-turboprop service. The company was formed in August 1971 and began services in February 1972.

Three 727-100s are operated in a simple colour scheme of orange cheatline below the windows and orange roof tiles plus stylised 'A' logo on the fin. The photograph shows 727-25, HK-2705X (c/n 18282) at Miami on 27 November 1983. This aircraft was bought from Eastern Air Lines in November 1981, then sold back to them immediately after this photograph was taken. (W.F. Wilson)

LACSA

COSTA RICA

Lineas Aereas Costarricenses SA (LACSA), Costa Rica's national carrier, was founded in 1945 by Pan American, the Costa Rican Government and Costa Rican private interests. Operations began in June 1946, and in 1952 the company took over Taca de Costa Rica, which had operated since 1939 as a unit of the Taca system. Today, LACSA operates regional passenger and cargo services from San Jose to Los Angeles, New Orleans, Miami, New York, San Juan (Puerto Rico) San Salvador, Panama City, Mexico City and Cancum.

Points also served are Guatemala City, San Pedro Sula, Cartagena, Baranquilla, Caracas and Maracaibo. LACSA currently operates five 727-200s of which three are leased. The colour scheme is a plain broad dark blue cheat line below the windows with red/white/blue striped fin and stylised dark blue titles.

The photograph shows 727-2Q6, N1280E (c/n 21972), named 'Bri Bri', taxying at Miami on 31 October 1989. (Henry Tenby)

STERLING AIRWAYS

DENMARK

Sterling Airways is wholly owned by Tjaereborg International Holdings, a leading Scandinavian travel agency and tour organiser. The airline was formed in May 1962 and began operations in July of that year. It operates inclusive tour flights to points in Europe, North Africa, North America and the Indian Ocean. The first Boeing 727 was delivered in November 1973 to supplement the airline's Caravelle fleet. Sterling Airways currently operates thirteen 727-200s, together with six Caravelles and its Boeing 757s. The colour scheme is red and white, with a red cheat line running under the windows and a sloping 'S' on the fin.

The photograph shows 727-2B7, OY-SBH (c/n 22164) in a variation of the normal scheme, on approach to Palma, Majorca in August 1987. (Jean Magendie)

ETHIOPIAN AIRLINES

ETHIOPIA

Ethiopian Airlines is the national carrier of Ethiopia. As well as a domestic network serving over forty points, Ethiopian Airlines operates international services linking Addis Ababa with points in Africa, the Middle East and Europe. The company was formed in December 1945. Scheduled services began in April 1946, with initial technical and managerial assistance from TWA. Routes were developed to neighbouring African and Red Sea Countries and in June 1958 Frankfurt became the first west European city to be

served, followed by Athens, Rome, Paris and London. Ethiopian currently operates three 727-200 series, the first of which was delivered in December 1981. The colour scheme comprises a yellow cheat line with red and green lining, as well as a red/yellow/green wing design on the fin. A yellow rampant lion appears on the nose and titles are in red.

The photograph shows 727-260, ET-AHK (c/n 22759), taxying in at Nairobi, Kenya on 22 March 1986 (Adrian Balch)

TAME

ECUADOR

TAME (Transportes Aereos Militares Ecuatorionos) was formed in 1962 and is a branch of the Ecuadorean Air Force. It began operations with Douglas DC-3 aircraft. Destinations initially served included Quito, Ibarra, Riobamba and Tulcan. A government decree legally created the airline as Transportes Aereos Militares Ecuatorianos on 20 May 1964. Scheduled passenger and cargo services are operated, plus charter services, linking points throughout Ecuador and the Galapagos Islands. TAME currently operates three 727-100s and one 727-200. The colour scheme features a smart dark blue, gold and medium blue cheat line on an otherwise white fuselage. The fin is medium blue with dark blue and gold trim. The airline's bird insignia appears in a circle in white on the fin.

The photograph shows the airline's sole 727-200, HC-BHM (c/n 22078) named 'Colopaxi' on a pre-delivery test flight over Seattle in September 1980. (Boeing)

AIR FRANCE

FRANCE

Air France was founded on 30 August 1933, when Societe Central pour L'Exploitation de Lignes Aeriennes purchased the assets of Compagnie Generale Aeropostale. Air transport was nationalised after World War Two and Societe Nationale Air France was set up on 1 January 1946. This was followed by Compagnie Nationale Air France on 16 June 1948, when the airline was incorporated by Act of Parliament. Today, the company operates an extensive network of routes stretching worldwide. Air France serves cities throughout Europe, North Africa, the Middle East, North and South America, the Caribbean, Africa, Madagascar and the Indian Ocean, as well as operating routes to the Far East.

The individual Air France orders for the 727 were not large, and were placed without fanfare, as they represented a choice of the American aircraft over the stretched Super Caravelle. Indeed, the first order was announced simultaneously with an order for three Caravelle IIIs. The 727s did not replace the Caravelles, more especially on the shorter denser routes, although work was found for most of the displaced Caravelles in the expansion years of the late sixties. The London – Paris route had grown to the point where a wet leased Modern Air Transport Convair 990 was being operated on the route. It was, therefore, the first route for the 727s, which began service on it on 23 April 1968. Other early 727 routes included Paris – Nice, and, in 1970, the thrice weekly non-stop Paris – Moscow route. Although the Airbus A.300 has since replaced the 727s on some flights on the denser routes, all seventeen Air France 727-228s are still in service. Their original colour scheme was natural metal with a white top and dark blue cheat line and fin design. In the mid-70s, this gave way to the current scheme which has an all-white fuselage devoid of cheat line. Blue/white/red national colours appear on the fin in a series of stripes and titles are in blue.

The photograph shows 727-228, F-GCDC (c/n 22083) over Washington, prior to delivery in April 1980. (Boeing)

AIR CHARTER INTERNATIONAL

FRANCE

Air Charter operates worldwide charters from thirty-five French provincial cities. Main destinations are Greece, Spain, Tunisia, Italy, Morocco, Turkey, Israel, Egypt, Canada and the USA. The company was formed in 1966 as a wholly-owned subsidiary of Air France. In 1978, Air Inter was granted twenty per cent of the shares, in return for agreeing not to operate charter services on the flag-carrier's routes.

Initially using ex-Air France Caravelles, Air Charter International acquired Pacific Southwest's two oldest 727-200s in 1972 and now operates seven of the type in its all-jet fleet. The revised colour scheme is white overall, with thin blue and red stripes running down the fuselage above the windows. The stripes extend up the fin and thicken in the form of a flying flag.

The photograph shows 727-2X3, F-GCMV (c/n 22608) in Air Charter International's old livery on a pre-delivery test flight in March 1981. (Boeing)

LUFTHANSA

GERMANY

Lufthansa German Airlines (Deutsche Lufthansa AG), one of Europe's largest carriers, operates an extensive worldwide system of scheduled passenger and cargo services to 178 points in eighty-four countries in Europe, Africa, the Near, Middle and Far East, Australia, New Zealand and North and South America. Operations to the former East Germany linking Frankfurt and Leipzig, commenced on 10 August 1989. Lufthansa is sixty-five per cent Government owned.

The airline was originally founded in January 1926, although its winged crest insignia originated with Deutsche Luftrederei in 1919. The company was reformed as Luftag on 6 January 1953, adopted its present title in August 1954 and began operations in April 1955.

The first European 727 service was operated by Lufthansa on 16 April 1964 with one aircraft operating Hamburg-London-Dusseldorf and vice-versa. On 10 May 1964, service from Frankfurt to London and Frankfurt to Istanbul and Beirut began, followed by Baghdad and Tehran on 25 May 1964. By the end of 1964, several European cities were being served by six 727s. The 727s had been ordered to replace the last piston engined aircraft and the 720Bs, which were proving uneconomic on the short runs. The 727 fleet was to be reassigned to the longer European and Near East routes when the short range jets such as the 737s were delivered. By 1967, all the

East European capitals had been added to the 727 network. The 727QCs enabled Lufthansa to dispense with leased Capitol C-46s on cargo flights.

Although the first 727-200s were not advanced models, they were equipped with the wide body interior. In 1973, eight 727-200s were ordered to replace the eight oldest 727s on the high density European routes. Today, Lufthansa has fifty-six 727-200s in service.

The original colour scheme was natural metal with white roof and fin. Cheat line was dark blue, as were the roof titles in capital letters. The blue and yellow Lufthansa insignia was presented as a flag on the fin. This scheme was modified in the early '70s with an all-dark blue fin upon which the Lufthansa insignia appeared on a yellow disc. The roof titles were changed to lower-case lettering. In the late '80s, the livery changed once again to the current scheme. The cheat line disappeared and the fuselage was painted white overall. However, by 1992 the Boeing 727 was the only type yet to appear in this scheme, which indicates the type's imminent withdrawal from Lufthansa service.

The final Lufthansa scheme for the type is shown in the photograph of 727-230, D-ABHI (c/n 205600), which has been in service since May 1972 (Deutsche Lufthansa).

CONDOR FLUGDIENST GERMANY

Condor Flugdienst is a wholly-owned subsidiary of Lufthansa. The carrier concentrates on charter and inclusive-tour operations with emphasis on a high quality of service. Its 727s have since been retired and all other Boeing aircraft transferred to a new Lufthansa subsidiary, Sudflug. Condor was formed in 1961 by the merger of Deutsche Flugdienst (founded in 1955 as a Lufthansa subsidiary) and Condor Luftreederei (founded in 1957 and acquired by Lufthansa in 1959).

The colour scheme on Condor's 727s was the same as that for Lufthansa, except the fin was all yellow instead of blue. Condor roof titles and tail insignia were in black.

The photograph shows 727-230, D-ABPI (c/n 20677) at Palma, Majorca on 26 March 1976. This aircraft spent much of its time between Condor and Lufthansa. (Adrian Balch)

HAPAG LLOYD

GERMANY

Hapag-Lloyd Fluggesellschaft, a German charter carrier, is a wholly owned subsidiary of the Hapag-Lloyd shipping group. The company operates inclusive-tour and other types of charters to the Mediterranean, the Canary Islands, central and eastern Europe and West Africa. The company was formed in July 1972 as Hapag-Lloyd Flugzeug and operations began in 1973. The company received federal authority to merge with Bavaria-German Air in January 1979. Hapag-Lloyd purchased its first three 727s in 1973 from All Nippon Airways, which were converted by Boeing with a new 125 passenger interior. The three aircraft had become five by the 1979 season and increased to eight in 1976. However, Hapag-Lloyd currently operate only two 727s. The colour scheme is natural metal with white top and fuselage sides. Cheat line is orange and dark blue with an orange fin upon which is the airline's stylised 'H' logo.

The photograph shows 727-81, D-AHLN (c/n 18952) in lovely winter sunlight and black sky at Hamburg in December 1985. (Michael Roser)

JETAIR

GERMANY

Jetair was a West German charter airline that flew international services from its base at Munich. The airline commenced operations with two 727-100s in 1984 until it ceased operations in 1985. The colour scheme was very simple, being white overall broken by a blue sash on the nose and centre engine. Jetair titles appeared on the fin in black.

The photograph shows 727-81, D-AJAA (c/n 18951) parked at Bournemouth-Hurn on 30 May 1986 after the airline ceased operations. (Adrian Balch)

AVIATECA

GUATEMALA

Aerolineas de Guatemala-Aviateca, known until 1974 as Aviateca Empresa Guatemalteca de Aviacion, operates scheduled services from Guatemala City to Houston, New Orleans, Miami and Flores, its one domestic route. The company was privatised in 1989, the Guatemalan Government holding being reduced; additional shareholders include TACA or El Salvador, Aviateca employees and other private investors. The company was established as the state airline in March 1945, to succeed Aerovias de Guatemala, founded in 1939. Aviateca began with domestic routes, using DC-3s and acquired a DC-6B in 1961. Jet services began with a leased One-Eleven 500 in November 1970. Aviateca operated two 727-100s from 1979 until the end of the 1980s. The airline has had varied colour schemes, but the 727s were natural metal overall with broad red/white/yellow cheat lines that sweep up the fin. Titles were in white on the red cheat line and the company's insignia appeared on the nose. The photograph shows 727-173C (c/n 19506) taxying at Miami in November 1987. (Author's collection)

SAHSA

HONDURAS

Servicio Aereo de Honduras (SAHSA) operates scheduled passenger services from Tegucigalpa to Belize, Guatemala, San Salavador (El Salvador), Managua (Nicaragua), San Jose (Costa Rica), Panama, San Andres Island (Colombia) and points in southern USA. The company, originally formed by Pan American Airways, but now owned by local interests, has been in existence for over forty years, having operated its first service on 22 October 1945 using a Douglas DC-3.

The five blue stars of the Honduran national flag occupy a white section of the fin which is a continuation of the dramatic cabin roof sweep, applied to an all-blue base colour. Between the distinctive coloured roof and a lower white band runs a narrow green cheatline which, whilst underlining the black 'SAHSA' titles, also contains tiny 'Servicio Aereo de Honduras SA' subtitles. SAHSA have only ever owned one 727-100, although a second was leased during the 1980s.

The photograph shows 727-81, HR-SHE (c/n 18823) which SAHSA have operated since December 1981. It is seen taxying out at Miami on Christmas Eve 1985 (W.F. Wilson)

ICELANDAIR

ICELAND

Icelandair (Flugleidir RF), the Icelandic flag carrier, operates international and domestic scheduled and charter flights. Icelandair was set up in its present form on 1 August 1973, as the holding company for Flugfelag Islands (the original Icelandair, founded in 1937) and Loftleidir (Icelandic Airlines, formed in 1944). The company assumed all operating responsibility in October 1979. The first 727 was ordered in August 1966 and entered service on 3 July 1967 on the Reykjavik – Glasgow – Copenhagen route replacing a Viscount. A second 727C was purchased in 1971 from Grant Aviation Leasing to replace the last Viscount. Today, Icelandair only owns one 727-200 which is on lease to Alaska Airlines. Colour scheme is natural metal with white roof and fin. A simple blue cheat line runs through the windows with a narrow line below the windows on a white band. The airline's symbol on the fin comprises what appears to be a twin flag in blue, which probably symbolises the merger of the two airlines, Icelandair and Loftleidir.

The photograph shows 727-208, TF-FLI (c/n 22295), which is the sole remaining 727 owned by the airline, at London Heathrow Airport on 25 April 1981 (W.F. Wilson)

IRAQI AIRWAYS   IRAQ

Iraqi Airways operates scheduled passenger and cargo services from Baghdad to points in the Middle East, Europe, Asia and North Africa. A domestic network is also operated. The airline was formed in 1948 as a Government-owned organisation and began domestic operations between Baghdad and Basrah on 29 January 1946 with Rapides. Tridents went into service on the London route in November 1965, the airline ordering three 727-200s on 30 January 1975 to replace these. Another three followed and all six are currently in service. Colour scheme comprises a medium green cheat line above which a dramatic section of dark green sweeps down the roof and up the fin. Titles are in white; on the roof in English and in Arabic on the fin. The airline's bird insignia appears in dark green on a white oval on the fin.

The photograph shows 727-270A, YI-AGM (c/n 21199) at London Heathrow Airport in June 1976, shortly after delivery. (Author's collection)

ALITALIA   ITALY

Alitalia (Linee Aeree Italiane), the Italian national airline, operates a worldwide network of scheduled passenger and cargo flights from Milan and Rome, serving Europe, North and South America, the Middle and Far East, north-east and South Africa and Australia. An extensive network of domestic services is also maintained. The airline was originally formed in September 1946 and operations began in May 1947. The present name was adopted in 1957 when the two major Italian airlines, Alitalia and LAI, merged. Alitalia carefully evaluated the 727 when N7003U stopped at Rome from 18-20 September 1963. In the event, Alitalia ordered a large fleet of DC-9s. However, Alitalia subsequently purchased eleven 727-200s, which commenced delivery in 1976, but these were only interim equipment until Airbus A300s were delivered and all Alitalia's 727s were sold by 1984. Colour scheme is a broad holly green cheat line on an all-white fuselage. The white fin is surmounted by a red/white/green 'A' design.

The photograph shows 727-243, I-DIRI (c/n 21265) overflying the mountains of Washington on a pre-delivery test flight in December 1976. Named 'Citta di Siena', this aircraft was sold to People Express in July 1984 and currently flies with Continental Airlines. (Boeing)

AIR AFRIQUE

IVORY COAST

Air Afrique is the international carrier of ten independent African countries. Based at Abidjan, Ivory Coast, the company operates an extensive network of regional services in Africa and serves a number of destinations in Europe. New York and Jeddah are also included in the carrier's international network. At present, the airline is seventy-two per cent owned by the following states: Benin, Burkino Faso, Central African Republic, Chad, Congo, Ivory Coast, Mauritania, Niger, Senegal and Togo. Air Afrique was officially formed in March 1961 by an agreement between eleven independent African states, formerly French colonies, and the Societe pour le Development du Transport Aerien en Afrique (Sodetraf).

Operations began in August 1961. Air Afrique has never owned any 727s, but leased a couple from JAT-Yugoslav Airlines for three years in the early 1980s. Colour scheme is a two-tone green cheat line through and below the windows. The airline's insignia appears on the fin in dark green.

The photograph shows 727-2H9, YU-AKK (c/n 22665), one of two leased from JAT in full Air Afrique colours, taxying at Schiphol in July 1986. (Author's collection)

AIR JAMAICA   JAMAICA

Air Jamaica was formed in October 1968 by the Jamaican Government and Air Canada. Services began on 1 April 1969 over a New York-Miami route. The carrier took over Air Jamaica Limited, a joint government venture with BOAC and BWIA which had operated Kingston-New York services with leased aircraft since 1965. Today, scheduled passenger and cargo services link the Jamaican capital and Montego Bay, with San Juan, Port au Prince, Grand Cayman, Miami, New York, Philadephia, Baltimore, Atlanta, Los Angeles and Tampa. Direct services to London are maintained in conjunction with British Airways. Late in 1974, Air Jamaica leased two 727-200s from National Airways Leasing then ordered two in its own right. These were followed by another two ordered in 1975, enabling the two leased aircraft to be returned. Colour scheme is quite dramatic with a yellow roof, broad orange cheat line under which is a narrow mauve cheat line. 'Air Jamaica' titles are carried on the lower fuselage in lower-case lettering. The colours sweep up the fin to the airline's bird insignia, which is yellow on an orange disc.

The photograph shows 727-2JO, VR-CMD (c/n 21108) at Miami in February 1980. (Nigel Chalcraft)

ALL NIPPON AIRWAYS   JAPAN

All Nippon Airways (ANA) was formed in December 1952 as Japan Helicopter and Airplane and scheduled operations began in 1953. JHA was merged with Far East Airlines in 1958 and took over Fujita Airlines in 1963, Nakanihon Air Services in 1965 and Nagasaki Airways in 1967. Today, ANA is one of Japan's three major airlines, which launched its first scheduled international passenger service in March 1986, linking Tokyo with Guam. Before this, the Japanese flag carrier (JAL) had a monopoly of all international routes. Subsequently, the carrier added Los Angeles, Washington, Beijing, Hong Kong, Seoul and Sydney to its international network. In 1989 new international services were added, including Tokyo-London (Gatwick). In addition to its international services, ANA operates scheduled passenger and cargo services on over seventy domestic and regional routes, as well as limited short-haul charter services.

Three 727s were ordered early in 1964, and in the meantime, a leased Boeing-owned 727 was used to begin service, 15 May 1964 between Tokyo and Sapporo. To meet traffic demands, All Nippon ordered a fleet of 727-200s and was the first airiine to put the Advanced 727-200 in service. The first of ANA's fleet of eleven 727-200s were ordered in July 1970 and served until all were sold in 1991. Final colour scheme was a two-tone blue cheat line that ran from a point on the nose, becoming thicker until it covered the fin. Titles in Japanese appeared on the roof in black and English 'ANA' up the fin in white.

The photograph shows 727-281, JA8344 (c/n 50573) in ANA's final scheme on the type, at Tokyo in December 1985. (Author's collection)

KOREAN AIR LINES   KOREA (REPUBLIC OF)

Korean Air operates an extensive international network of scheduled passenger and cargo services linking Seoul with Frankfurt, London, Paris and Zurich in Europe; Los Angeles, New York, Toronto and Vancouver in North America; Bahrain, Jeddah and Riyadh in the Middle East; and Bangkok, Fukuoka, Hong Kong, Kuala Lumpur, Manila, Nagasaki, Nagoya, Nigata, Osaka, Sapporo, Singapore, Taipei and Tokyo in Asia. In addition, domestic services link Seoul with Cheju and Pusan. A non-stop service to Europe started on 25 March 1990. KAL acquired three 727-100s from Japan Air Lines in 1980, which it had leased since 1972. Subsequently, KAL bought All Nippon Airways' fleet of 727-200s as they became available. The original colour scheme for these aircraft was a traditional blue cheat line with a thin red-on-white cheat line below it. Red trim and company logo appeared on the fin. This has since been replaced by the entire upper fuselage and fin being painted pale blue, with a grey cheat line down the windows. The airline has also changed its titles from 'Korean Air Lines' to 'Korean Air' to reflect the radio callsign.

The photograph shows 727-281, HL7348 (c/n 20435) in KAL's old colour scheme in August 1984. This aircraft has been in service since February 1980. (Author's collection)

KUWAIT AIRWAYS

KUWAIT

Kuwait Airways was formed in March 1954 under the title Kuwait National Airways Company and became the Kuwait Airways Company in March 1957. BOAC (now British Airways) took over technical management under a five-year contract in May 1958 and British International Airlines was absorbed in April 1959. The airline became wholly Kuwaiti Government owned in 1962. As a result of the 1991 Gulf War, Kuwait Airways operated reduced services from Cairo, as part of its fleet has been seized by Iraq. The country's national carrier is now recovering, with new equipment on order, and provides scheduled passenger and cargo services from Kuwait to forty-one cities in thirty-eight countries in the Middle and Far East, Europe, North Africa and North America. Three of the airline's four 727-200s survived the Gulf War and are currently in service. Colour scheme features a broad windowline and tail band in ocean blue, trimmed on both sides in black, providing a bright contrast to the white fuselage top. The company's bird symbol flies on the tailband, reversed out in white. Kuwait Airways titles are applied in blue on the cabin roof in both English and Arabic.

The photograph shows 727-269, 9K-AFA (c/n 22359) named 'Warba', which has been in service since September 1980, at Dubai in December 1984. (Author's collection).

LIBYAN ARAB AIRLINES

LIBYA

The national airline of the Socialist People's Libyan Arab Jamahiriya, was formed in 1964 by the Libyan Government and commenced domestic passenger services in October 1965 as Kingdom of Libya Airlines. The present title was adopted following Libya's revolution in September 1969 and today the company provides domestic scheduled services connecting Tripoli with nine major towns. International routes link Tripoli, Benghazi and Sebha with Algiers, Athens, Belgrade, Cairo, Frankfurt, Madrid, Moscow, Paris, Rome, Prague, Warsaw and Zurich. Services to London are currently suspended. Libyan Arab Airlines purchased two 727s which had been built against a Continental option which was not taken up. With technical assistance from Air France, which maintained the aircraft and provided flight crews, the 727s were introduced on the routes from Tripoli and Benghazi to Cairo, Beirut and other

European and Middle East cities, supplementing Caravelles. One of the 727s was shot down over the Sinai desert on 21 February 1973 by the Israeli Air Force with the loss of 106 lives, but nine remain in service today. The colour scheme is a blend of gold and chocolate brown with white upper fuselage. A wide gold cheat line is trimmed with brown pinstripes, as is the base of the golden tailfin. The company motif, representing a gazelle in full flight, appears in chocolate brown on a white disk. Libyan Arab Airlines titles are presented in English on the starboard side and in Arabic on the port.

The photograph shows 727-2L9, 5A-DIB (c/n 21051) on finals to London Heathrow on 26 August 1983, when services there were still permitted (Adrian Balch)

AIR MALI

MALI

Air Mali was formed in October 1960, as the national airline of the Republic of Mali. Operations began in 1961 with technical and equipment assistance being provided by Aeroflot. The company was helped by a gift from the British Government of three ex-BEA DC-3s. Scheduled passenger services are operated to various domestic points, as well as international cities in France, Ivory Coast, Nigeria, Algeria, Morocco and the Congo Republic. Air Mali operated a leased Airlift 727 from March 1971 and a second aircraft

leased from World in July 1971, which was eventually purchased in 1974. The 727s were employed on services to Marseilles and Paris, replacing Il-18s. Colour scheme comprised a yellow cheat line trimmed with dark green. Company titles and tail flash were also dark green.

The photograph shows 727-173C, TZ-ADR (c/n 19509) at Stansted on 25 March 1979. Air Mali ceased operations in 1987 and this aircraft was sold to Federal Express. (W.F. Wilson)

ROYAL AIR MAROC

MOROCCO

Royal Air Maroc was formed in 1953 as Compagnie Cherifienne de Transports Aeriens. This was a result of a merger involving Air Atlas and Air Maroc. The airline became the national flag carrier of newly-independent Morocco on 28 June 1957. RAM ownership is controlled by the government, Air France and private interests. The company has financial holdings in Royal Air Inter. Scheduled passenger and cargo services are operated from Casablanca to points in Morocco, North and West Africa, Europe, the Middle East, New York and Montreal, as well as to Rio de Janeiro. Eight 727-200s are currently operated in the fleet. Colour scheme features a green, white and red cheat line with white upper surfaces and grey undersides. Royal Air Maroc titles appear in lower case red lettering on the starboard side in English and on the port side in Arabic. The centre piece of the national flag is the green pentangle (or Seal of Solomon), which adorns the fin in the form of a shooting star, whose red tail encompasses the red RAM initials.

The photograph shows 727-2B6, CC-CCG (c/n 20471) on finals to London Heathrow Airport on 21 September 1989. (Adrian Balch)

AIR NAURU

NAURU

The tiny island republic of Nauru, in the central Pacific, has a total land area of less than eight square miles, but nevertheless boasted its own all-jet national airline until its Air Services Certificate was suspended on 24 June 1988. However, services were restored with an all-Boeing 737 fleet. Formed in February 1970 by the Nauru Government as a division of the Directorate of Civil Aviation, Air Nauru flies scheduled passenger and cargo services to some two dozen regional destinations. Main routes are those which link the islands with Melbourne and Sydney (Australia), Auckland (New Zealand), Singapore, Hong Kong, Taipei (Taiwan) and Okinawa and

Kagoshima (Japan). Colour scheme comprises a dark blue wedge cheat line, which has yellow through its centre. The fin is dark blue as an extension of the cheat line and displays a large twelve-pointed star, from the flag of Nauru, beneath a yellow horizontal bar. Air Nauru titles are also in dark blue alongside the airline's 'V' motif on the cabin roof.

Although no longer in service, the photograph shows 727-95, C2-RN5 (c/n 19252), one of three once operated by the airline, at Kai Tak Airport, Hong Kong on 14 June 1980. This aircraft was sold to Evergreen International Airlines in April 1985. (Adrian Balch)

ROYAL NEPAL AIRLINES

NEPAL

Royal Nepal Airlines was formed in July 1958 to take over domestic services from Indian Airlines, which had been operating within Nepal for some eight years. Today, Kathmandu is linked by regular passenger services, with numerous outlying communities throughout Nepal. International routes fly regionally to New Delhi, Calutta and Patna (India), Dhaka (Bangladesh), Hong Kong, Bangkok (Thailand), Karachi (Pakistan), Colombo (Sir Lanka), Singapore and Dubai (UAE). Royal Nepal Airlines purchased their first 727 in June 1972, which was shortly joined by a second, both remaining in service to this day. The original colour scheme was a red and navy cheat line

with Royal Nepal Airlines titles in English capitals on the port side of the white roof, and Nepalese on the starboard side. The flag of Nepal was featured on the white fin. Since 1986, this scheme has been changed to an overall white fuselage with just a diagonal band of red and dark blue up the fin and titles as before, but in lower case lettering preceeded by a small Nepalese flag.

The photograph shows 727-1F8, 9N-ABD (c/n 20421) in the airline's old colours, on a pre-delivery test flight over Washington in August 1972. (Boeing)

AIR PANAMA

PANAMA

Air Panama International is designated as the flag-carrier of Panama and operates scheduled passenger and cargo services from Panama City to Mexico City, Bogota, Guayaquil, Lima, Caracas and Miami. Air Panama was formed in August 1967 with assistance from Iberia, and began operations in August 1967. Initial operations were with a DC-9, leased from McDonnell-Douglas. The first 727 replaced the DC-9 in November 1972 on routes from Panama City to Guatemala City and Mexico City to Miami, to Guyaquil and Lima and to Bogota. The airline had three 727-100s but now has only one. Colour scheme is natural metal overall, with broad red, white and blue cheat line, which sweeps up the fin. The company's 'AP' logo appears on a white disc on the fin and roof titles are in white-outlined red and blue.

The photograph shows the airline's sole 727-81, HP-619 (c/n 18920) taxying at Miami on 21st November 1983. (W.F. Wilson)

AERO PERU
PERU

Aero Peru (Empresa de Transporte Aereo del Peru), the country's flag-carrier, operates domestic services linking Lima with other points throughout Peru and North, Central and South America. The airline was formed in May 1973 as the wholly Government-owned national carrier of Peru. The company is the result of a reorganisation of Satco, the airline element of the Peruvian Air Force. Satco began operations in 1960 when Transportes Aereos Militares (founded in 1931 as Lineas Aereas Nacionales) was split into two segments – TANS to operate subsidised feeder services to points in the Selva region and Satco to operate trunk routes. AeroPeru was denationalised in July 1981, and in May 1974, purchased their first 727 from Eastern, which was used to begin international services in July 1974

to Buenos Aires, Santiago, Guyaquil and Miami. Colour scheme at the time was a startling pink cheat line with red cabin upper surfaces and fin, which must have been counterproductive in reflecting the sun. Since then, the colour scheme has been changed to a more conventional white cabin roof and fuselage sides, with two thin red cheat lines running above and below the windows. The fin remains red with the airline's sun motif in white. The rudder sports the country's red, white, red national colours.

The photograph shows, 727-25, OB-R-1081 (c/n 18269) in AeroPeru's old colour scheme at Lima in March 1976. This aircraft was withdrawn from use in September 1983 and used for spares. (Author's collection)

PHILIPPINE AIRLINES

PHILLIPPINES

Philippine Airlines celebrated its fiftieth anniversary in March 1991, as the longest-serving airline in Asia. It operated scheduled passenger and cargo services over a forty-three-point domestic network, together with international flights to Singapore, Kuala Lumpur, Taipei, Hong Kong, Kota Kinabalu, San Francisco, Los Angeles, Honolulu and Guam, Tokyo, Melbourne, Sydney, Brisbane, Ho Chi Minh City, Bandar Seri Begawan, Bangkok, Karachi, Rome, Frankfurt, Paris, Dhahran, Dubai, Riyadh and London. The airline is ninety-nine per cent owned by the Government. PAL was formed in March 1941 to operate scheduled services. Domestic operations were resumed after the war in February 1946. International operations started in 1946, were suspended in March 1954 and resumed in June 1962.

In July 1979, PAL leased two 727-200s from Hughes Airwest for two years for evaluation pending a decision on purchasing new equipment. Colour scheme was based on an all-white fuselage with thin red/white/blue stripes running underneath the windows. The fin was entirely red and dark blue, divided by a thin white line. Standard small lower-cases titles were in black on the roof with a small Philippine flag.

The photograph shows 727-2M7, N725RW (c/n 21502), in March 1980. After the lease this aircraft returned to Republic Airlines who had, meanwhile, merged with Hughes Airwest. Republic has since been absorbed by Northwest, with whom the aircraft flies today. (Author's collection)

TAP-AIR PORTUGAL

PORTUGAL

Air Portugal–TAP, the Portuguese national carrier, operates an extensive network of the scheduled passenger and cargo services to the destinations in North and South America, Africa and throughout Europe. Domestic flights serve Madeira, the Azores, Faro and Porto. Other destinations comprise Casablanca and Sao Tome, Newark and Curacao. Originally founded as Transportes Aereos Portugueses on 14 March 1945, as a Special Department of the Secretariat for Civil Aviation, TAP began operations in 1946. The airline was nationalised in 1975. TAP selected the 727 for its field performance, particularly for use at Funchal in Madeira. The principal routes for which the 727s were purchased were Lisbon to Oporto, to Faro, to Funchal and to Portuguese Guinea. Service to various European capitals, however, was also flown by 727s, replacing Caravelles. TAP ordered its first 727-100 in March 1966 and disposed of its last three 727-200s in 1991, which were replaced by Airbus A310s. Colour scheme was a mainly white fuselage with a broad band of red running beneath the windows, topped by a thin green line, making the Portuguese flag colours. The cheat lines ran up the fin culminating in the 'TAP' logo from Transportes Aereos Portugueses in red.

The photograph shows 727-282, CS-TBS (c/n 20973) named 'Lisboa' on finals to London Heathrow Airport on 7 May 1989. (Adrian Balch)

AIR ATLANTIS
PORTUGAL

Air Atlantis, the charter subsidiary of TAP-Air Portugal, operates regular year-round flights to Portugal from Austria, Belgium, Canary Islands, France, Germany, Holland, Ireland, Italy, Scandinavia, Spain and Switzerland. From 1990, Portugal to Cairo and Tunis routes were added. Ad hoc charters are also operated. The UK is a major market, with operations from Gatwick, Luton and fifteen regional airports. The carrier's main base is at Faro. Crew training, all support services and head offices resources are provided by TAP. Air Atlantis was set up in 1985 and operations began mid-1985. The colour scheme features a similar cheat line to that of TAP, but with green the dominant colour. A green and red 'A' logo appears on the white fin. Air Atlantis leased five of TAP's 727s during 1987-89, but they are no longer in service. The photograph shows 727-232, CS-TCI (c/n 20867) on finals to London Gatwick Airport in July 1988. (Martin Siegrist)

SINGAPORE AIRLINES SINGAPORE

SIA was formed on 28 January 1972, as the wholly Government owned national airline of Singapore to succeed the jointly operated Malaysia-Singapore Airlines. Operations began on 1 October 1972. Today, daily flights to thirty-six countries link sixty-two cities in Europe, the Middle East, Asia, South West Pacific, Australia/New Zealand and North America. European destinations include London, Manchester, Berlin, Brussels, Paris, Copenhagen, Amsterdam, Frankfurt, Zurich, Vienna, Rome, Athens and Istanbul. Honolulu, Vancouver, San Francisco, Los Angeles and Dallas/Fort Worth are

cities served in North America. SIA operated up to six 727-200s during 1978-83, but all have since been sold. Colour scheme was based on an overall-white fuselage with a yellow and dark blue cheatline in a wedge-shape running from the front cabin window to the rear. The whole fin was dark blue upon which SIA's flying bird motif appears in yellow.

The photograph shows 727-212, 9V-SGJ (c/n 21948), seen taxying in at Kai Tak Airport, Hong Kong on 16 February 1980. This aircraft was sold to Alaska Airlines in March 1985. (Adrian Balch)

SOUTH AFRICAN AIRWAYS SOUTH AFRICA

South African Airways operates a domestic network of scheduled passenger and cargo services together with regional flights to several African cities. Intercontinental services link Johannesburg, Durban and Cape Town with Rio de Janeiro, Hong Kong, Ilha do Sal, Lisbon, Taipei, Rome, Vienna, Zurich, Frankfurt, Paris, Amsterdam, London, Manchester and Tel Aviv. SAA was formed on 1 February 1934, when the Union Government took over the assets and liabilities of Union Airways. The airline is owned and controlled by the Government of South Africa and falls under the jurisdiction of Transnet. South African 727s entered service on 1 August 1965 on the Johannesburg–Cape Town–Port Elizabeth–East London–Durban route, with all five of the first order entering service by 1 November 1965. The ninety-eight-seat 727s replaced fifty-six seat Viscounts,

which in turn replaced DC-7Bs and DC-4s on 'coach' services. The 727s were also employed on regional services to Bulawayo, Salisbury and Lourenco Marques. All SAA's 727s were sold by 1982. Final colour scheme on SAA's comprised a dark blue cheatline, trimmed with white and orange underneath. The fin was orange accompanied by SAA's winged Springbok insignia in dark blue, which also appeared alongside the roof lettering. The latter was in English on one side and in Afrikaans on the other.

The photograph shows 727-44, ZS-EKW (c/n 19318) named 'Komati', seen prior to delivery in January 1967. This aircraft was sold in December 1981. (Boeing)

IBERIA

SPAIN

Iberia (Lineas Aereas de Espana), the national airline of Spain, has a large network of scheduled passenger and cargo services to North, Central and South America, Africa, Europe, the Middle East and the Far East. An extensive domestic network is also provided. Iberia was founded in 1927 as Lineas Aereas de Espana, from a succession of carriers dating back to 1921. The carrier is state-owned and is controlled through the Instituto Nacional de Industria. Iberia became the fifty-ninth customer for the 727 in January 1972, and proceeded to order a total of twenty-nine aircraft. A subsequent order brought the total to thirty-five and all are currently in service. Colour scheme is overall white upon which a triple cheatline of red, orange and yellow sweeps over the top of the cabin roof and along the top of the cabin windows. White italic IBERIA titles are set into the red and orange bands of the cheatline. A cleverly quartered 'IB' logo in red and yellow on the white tail carries a royal crown.

The photograph shows 727-256, EC-CID (c/n 20974) named 'Malaga' on finals to London Heathrow Airport on 11 March 1978. (John Hughes)

TRANSAIR

SWEDEN

Transair Sweden AB was formed in 1950 as Nordisk Aerotransport AB to fly newspaper delivery flights and took its present title in 1951. From 1953, charters and later inclusive tour flights were operated, and C-46s and DC-6s were also operated in the Congo in the early 1960s for the United Nations and the Congolese Government. Night mail and passenger flights between Copenhagen and Malmo, and freight services from Copenhagen to Stockholm, Hamburg, Amsterdam, and Paris were operated on behalf of SAS. Transair Sweden became a subsidiary of Svenska Handelbanken in October 1968 and the first Boeing 727 was delivered at the end of 1967, eventually replacing the DC-7Bs then in use. The airline became a wholly-

owned subsidiary of SAS on 16 May 1975. The three 727s were flown on charters and inclusive tours for Scanair A/S, who operated the aircraft until they were sold to Philippine Airlines in October 1981.

Colour scheme was natural metal with white upper surfaces and a thin orange cheat line above the windows, running up the fin. An orange sun symbol appeared on the white fin, accompanied by the legend 'Sunjet'. Company titles appear on the roof in lower case black lettering.

The photograph shows 727-134, SE-DDA (c/n 19691) taxying on to the runway at Copenhagen/Kastrup Airport in March 1975. (Author's collection)

SYRIANAIR
SYRIA

Syrian Arab Airlines (Syrianair), the Government-owned national airline, opeated an extensive network of scheduled passenger and cargo services to Aleppo, Latakia and Deirezzor in Syria and to major cities in Europe, North Africa and the Middle and Far East. The airline was formed in 1961 to succeed Syrian Airways (formed in 1946). This led to the formation of United Arab Airlines by merging Misrair of Egypt and Syrian Airways, Syrian Arab's predecessor, which became independent again after the break with Egypt and took back its fleet and routes from UAA. Syrian Airways had operated domestic and regional routes from Damascus; Syrian Arab inherited these and started operating into Europe with DC-6Bs, serving Paris and London from 1964 and also flying east through the Gulf states to Karachi and Delhi. The first 727 was delivered in March 1976 to supplement Caravelle services. The colour scheme included a standard white cabin roof, with a medium blue cheatline through the windows trimmed by thinner lines in the same colour. The fin colour is the same blue with the airline's motif in white. Syrianair titles in both English and Arabic appear on the cabin roof in blue. Three 727-200s were delivered to Syrianair and all are currently in service.

The photograph shows 727-294, YK-AGC (c/n 21205) on a pre-delivery test flight near Seattle in April 1976. (Boeing)

CHINA AIRLINES

TAIWAN

China Airlines (CAL), the national airline of Taiwan, operated scheduled international passenger and cargo services from Taipei to Hong Kong, Bangkok, Phuket, Manila, Nagoya, Fukuoka, Singapore, Kuala Lumpur, Dhahran, Dubai, Jakarta, Denpasar, Tokyo, Okinawa, Seoul, Honolulu, Los Angeles, New York, San Francisco, Dallas, Anchorage, Vancouver, Luxembourg and Amsterdam. The carrier was formed in 1959 and became the national airline of Taiwan in 1965.

An L.1049H Super Constellation was used temporarily to begin scheduled international services on 2 December 1966 from Taipei to Saigon. The 727 replaced the L.1049 on 9 April 1967, and service from Taipei to Hong Kong, Osaka and Tokyo was also commenced.

The second aircraft was delivered in September 1967 and was used to expand services to include Seoul, Manila, Bangkok, Singapore and Kuala Lumpur. The third aircraft was ordered for increased services to all the above-mentioned points. All three served until late 1982, when they were transferred to the Republic of China Air Force. Colour scheme was a straightforward blue, white and red cheat line, which ran up the white fin. Taiwan's flag appeared on the fin and 'China Airlines' titles appeared on the white cabin roof in dark blue English and Chinese writing.

The photograph shows 727-109, B-1820 (c/n 19520) taxiying onto the runway at Kai Tak Airport, Hong Kong on 17 October 1980. (Adrian Balch)

TUNIS AIR

TUNISIA

Tunis Air (Societe Tunisienne de L'Air) was founded in 1948 by the Tunisian Government, which holds 45.2 per cent of the shares. Air France holds 5.6 per cent and other interests 49.2 per cent. Passenger and cargo services are operated from Tunis, Djerba, Monastir, Tozeur and Sfax to Abu Dhabi, Algeria, Austria, Belgium, Denmark, Dubai, Egypt, France, Great Britain, Greece, Holland, Italy, Jordan, Kuwait, Libya, Luxembourg, Morocco, Saudi Arabia, Senegal, Spain, Sudan, Switzerland, Syria and Germany. Tunis Air has eight 727-200s currently in service, of which the first was ordered on 10 March 1971. Colour scheme is based on a traditional white topped fuselage with grey undersides. The cheat line comprises two broad red bands through and below the windows, divided by a thin white line. A cleverly-designed logo appears on the tail in red, comprising an integrated 'ta' with the company's Gazelle motif in white.

The photograph shows 727-2H3, TS-JHT (c/n 21235) named 'Sidi Bou Said', over France in September 1988. (Jean F. Lipka)

TURK HAVA YOLLARI – TURKISH AIRLINES TURKEY

Turk Hava Yollari (THY Turkish Airlines) operates scheduled passenger services connecting Ankara, Istanbul, Antalya, Izmir and Adana with eleven other domestic points. In addition, an extensive network of international scheduled services is operated to Europe, the Middle and Far East, North Africa and the USA. The airline is owned by the Turkish Government. THY was formed in May 1953 as Devlet Hava Yollari. The present title was adopted in 1956, when the airline became a corportation.

The airline placed its order for 727s in 1973 and currently operates all nine of the type. Colour scheme is based on a white and grey fuselage with an attractive cheat line comprising multiple red and white stripes. The fin has slightly thicker red and white horizontal stripes, encompassing the airline's red sail insignia on a white disc. This colour scheme is currently being replaced by an overall plain white fuselage with red fin. Roof titles comprise red 'THY' and dark blue 'TURKISH AIRLINES'.

The photograph shows 727-2F2, TC-JBH (c/n 20982), named 'Antayla', seen on a pre-delivery test flight in November 1974. This aircraft subsequently crashed on Mount Karatepe on 19 September 1976. (Boeing)

AVENSA (AEROVIAS VENEZOLANAS) VENEZUELA

AVENSA operated domestic scheduled passenger services from Caracas to various points throughout Venezuela. International services are operated to Mexico City and Panama City. The airline was formed in June 1943, with domestic services starting in 1944, followed by international flights in 1955. In April 1976, the Venezuelan Government financial agency, Corporacion Venezolana de Fomento, bought the thirty per cent stock held by Pan American since the airline's foundation. The then international services of Avensa and LAV were merged in 1961 to form Viasa, in which Avensa has a forty-five per cent holding. Avensa is now wholly owned by the Venezuelan Government. The airline currently operates nineteen 727s of both variants, purchased from United, Braniff and other airlines. Colour scheme is overall white with huge 'AVENSA' titles in blue down the fuselage, through the windows. This style may be influenced by Pan Am's final scheme. On the fin, a winged map of South America appears in blue and white. The photograph shows 727-22, YV-87C (c/n 18853) at Miami in November 1987. This aircraft has been in service since March 1985 and was originally delivered to United Airlines in October 1965. (Author's collection)

AIR VIETNAM

VIETNAM

Air Vietnam was founded originally to take over the Air France services in the area. Air Vietnam was 92.75 per cent owned by the Government in 1975, shortly before the fall of Vietnam. The 727Cs were ordered for Air Vietnam by Pan American, after the latter carried out a technical survey in December 1967. These 727s were placed in service on the international routes to Hong Kong, Taipei, Bangkok, Kuala Lumpur and Singapore, replacing a Caravelle which was then used on internal services. Services were extended twice weekly from Taipei to Osaka and Tokyo from 30 July 1968. One 727 was destroyed in a war-related hijacking in 1974 and the other was in Taipei when South Vietnam fell on 29 April 1975. The colour scheme basically employed the Pan American blue cheat line through the windows with Air Viet Nam roof titles in matching blue. On an all-white fin, the airline's dragon head insignia was carried on a yellow disc.

The photoghraph shows 727-121C, XV-NJB (c/n 19818) at Osaka, Japan on 22 April 1970. Named 'Thanh Long' it served from January 1968 until siezed by the Taiwan Government in February 1976. After service with China Airlines, it is currently operated by the Republic of China Air Force. (Author's collection)

JAT-YUGOSLAV AIRLINES   YUGOSLAVIA

Jugoslovenski Aerotransport (JAT-Yugoslav Airlines) is a wholly state owned airline. An extensive network of scheduled passenger services is operated to sixteen domestic points and forty-eight cities in Western Europe, Scandinavia, Eastern Europe, North America, the Middle and Far East and Australia. Formed in 1946 by the government to take over interim services being flown with military aircraft, operations began in 1947 over domestic routes using Junkers JU-52/3ms, which were soon replaced by DC-3s. After Yugoslavia's break with the Soviet Union in 1948, JAT ceased operations temporarily, but resumed operations by early 1949. Since then, JAT has almost always managed to quietly resist pressure to buy Soviet-built equipment, and had operated Convair 440s, Caravelles and DC-9s before the 727s were introduced, supplementing DC-9s on the Belgrade to Zagreb and Belgrade to major European cities flights. JAT also acted as purchasing agent for two 727-2L8As which are operated by the Government and/or Air Force. Ordered in 1973-74, JAT currently operates eight 727-200 series. Colour scheme features a medium blue cheat line underlined by thin bands of white and red. The fin is mainly blue upon which the JAT red oval is placed with white lettering and surround. Titles 'JUGOSLOVENSKI AEROTRANSPORT' appear on the roof's port side in red and 'YUGOSLAV AIRLINES' on the starboard side. Recently, this scheme has been modified, with the cheat line being deleted and replaced by a blue, white, red and grey 'chin strap' which wraps round the nose.

The photograph shows 727-2H9, YU-AKF (c/n 21038), on finals to London Heathrow Airport on 13 June 1991. (Adrian Balch)

AVIOGENEX

YUGOSLAVIA

Aviogenex was formed in 1968 as Genex Airlines and is the airline division of the state import-export agency which is known as Generalexport. It concentrates on charter and inclusive tour flights, mostly within Europe, and operates four second-hand 727-200s alongside four 737s. Colour scheme pattern is the same as on JAT aircraft, with the blue being replaced by maroon and the thin red cheat line being repainted deep yellow. Roof titles are in black and the airlines insignia appears on the fin in white.

The photograph shows 727-2L8, YU-AKD (c/n 21040) named 'Zagreb' taxying at Manchester on 28 August 1985. (Paul Tomlin)

DAN-AIR SERVICES

UNITED KINGDOM

Dan-Air Services (Dan-Air London) is the airline subsidiary of Davies and Newman Holdings, London shipping brokers and travel group (established 1922), from which its name is derived. The company is a major operator of scheduled passenger services to international destinations. The airline is also a leading operator of inclusive-tour flights to resorts throughout Europe, the Mediterranean, the Canary Islands and the Middle East from more than a dozen airports in the U.K. under contract to most leading British tour operators. Dan-Air also operates contract services for oil companies, together with ad hoc passenger and cargo charters and night mail contracts. Dan-Air began operations in 1953 and currently operates the first 727s on the U.K. register. The first aircraft were three Japan Air Lines aircraft bought from the Norwich Union Assurance Group on long-term lease in 1973. When the Comet 4s were phased out of service, more second-hand 727s were purchased and twelve are currently on strength.

Colour scheme comprises a white and grey fuselage, upon which a red and blue cheat line starts at a point on the nose becoming thicker, until it sweeps up the fin. A Union Jack and black titles appear in the roof. The company's compass insignia and flag appear on a white disc on the fin.

The photograph shows 727-46, G-BAJW (c/n 18878) on finals to Manchester on 28 August 1985. This aircraft was bought from Japan Air Lines in December 1972. (Paul Tomlin)

AIR ATLANTA

U.S.A.

Air Atlanta was a privately-owned carrier established in May 1981. On 1 February 1984, the airline began scheduled services on Atlanta-Memphis and Atlanta-New York routes, later adding services to Philadelphia, Tampa, Miami and Orlando. On 3 April 1987, Air Atlanta suspended services and filed for Chapter 11 bankruptcy protection. The company was operating as part of the 'Pan Am Express' system. Colour scheme comprised a dark blue cheat line through the windows with two thinner cheat lines in two tones of red below the cheat lines, separated by thin white lines. Titles were in black and the white fin sported the company's insignia of curved red lines of varying thickness. The photograph shows 727-22, N1187Z (c/n 18323) at Miami in April 1987. This aircraft was bought from Korean Airlines in June 1984 by ATASCO Leasing Inc. It was leased to Air Atlanta until they ceased operations, then was leased to LADECO of Chile. (Andrew Brattkus)

ALASKA AIRLINES
U.S.A.

Alaska Airlines, a subsidiary of Alaska Air Group following its 1987 merger with Jet America, has established itself as the dominant carrier along the US West Coast. Originally founded as McGee Airways in 1932, Alaska Star Airlines became Alaska Airlines in 1944, following several mergers with smaller companies. Further mergers have occurred, the most recent with Jet America in 1987. The company operates a network of scheduled passenger and cargo routes linking California, Arizona, Idaho, Oregon, Washington and Alaska and four destinations in Mexico, Los Cabos, Mazatlan, Puerto Vallarta and Guadalajara. In June 1991, services began to the Soviet Far East cities of Magadan and Khabarovsk. Alaska Airlines currently operates twenty-eight 727s, alongside similar numbers of 737s and DC-9s. Colour scheme is white overall broken by huge 'Alaska' titles in dark blue script on the forward fuselage, accompanied by a thin double dark blue cheat line low on the fuselage. An Eskimo's face in dark blue and white dominates the fin.

The photograph shows 727-227, N298AS, (c/n 22004) seen taxying at Madeira, Ca. on 20 July 1991. (Stephen Wolf)

AMERICAN AIRLINES

U.S.A.

American Airlines was formed on 13 May 1934 as a direct successor to American Airways. Since the company's formation the carrier has played a vital role as a launch customer in the development of many civil aircraft. These include the Douglas DC-3, DC-7, CV-240, CV-990, Lockheed Electra and DC-10. American's huge route system extends from Bermuda to Hawaii, from Montreal and Toronto in the north, Acapulco and Mexico City in the south and across the Atlantic to Europe. The airline flies over twenty-five million people annually across its network, and with a fleet that is continually expanding, it is hardly surprising that the company ranks amongst the largest airlines in the world.

American's first order for 727s was placed as a letter of intent on 16 May 1961. American's first 727 operations were two daily New York (JFK) to Chicago round trips. Three more, including one from Newark, were added on 26 April 1964, and on the same day 727 services from Chicago to Dallas began. Today, American has no less than 125 727-200s and thirty-nine 727-100s in its fleet. Colour scheme is overall polished natural metal with a broad, blue, white and red cheat line. Cabin roof titles are white outlined red and the fin has the company's 'AA' letters in dark blue and red divided by a dark blue Eagle. The photograph shows 727-123, N1964 (c/n 19838) seen taxying in at Eppley Field, Omaha, Nb. on 17 October 1989. (Adrian Balch)

BRANIFF INTERNATIONAL
U.S.A.

Braniff Airways, a Dallas-based US major airline, was founded by two brothers, Paul and Tom Braniff, in 1928. Due to financial difficulties, the company was forced to close down on 12 May 1982. However, Braniff was able to recommence tailored-down operations in March 1984. Financial backing was given from Hyatt Inc., and Braniff operated a hub-and-spoke network of scheduled passenger services from Dallas/Fort Worth, to Boston, Chicago, Denver, Kansas City, Las Vegas, Phoenix, New York, San Francisco and Washington DC. Unfortunately, this limited service did not last long and once again Braniff Airlines filed for Chapter 11 protection on 28 September 1989, immediately reducing services from thirty-six cities to just eleven. All scheduled operations ceased on 6 November 1989. Braniff has had more experience of buying, selling and leasing 727s than anyone else. It has used both variants of the 727, the first being ordered in May 1965.

Braniff is most remembered for its dramatic colour schemes. When the 727s were introduced in 1966, Braniff had changed to a scheme where the whole fuselage/fin was painted in one of seven different colours, with white wings. This scheme was changed in the 1970s, to several two-tone schemes knows as 'Flying Colors'. During this period, a 727-200 and a DC-8 were repainted in a flamboyant unconventional scheme by the artist Alexander Calder. In the late 1970s Braniff changed its colour scheme yet again, still on the theme of several two-tone schemes, but using more subtle fuselage colours this time, with matching lighter-coloured strips on lower fuselage, on the fin and round the engines. This scheme lasted until the airline first ceased operations in 1984. When the airline was resurrected, a more conventional scheme was adopted with purple-blue under-fuselage divided by a light grey cheat line on lower fuselage and large red 'BRANIFF' titles on the forward fuselage and up the fin. This scheme was applied to Braniff's 727s, 737s and One-Elevens. In July 1991 Braniff International recommenced operations once again with limited services, taking over Emerald Air in the process.

The centre photograph shows 727-227, N456BN (c/n 21462) in bottle green with light green trim, the third multi-colour scheme for the airline's 727s.

This aircraft was delivered new in May 1978 and served until 1982. (Braniff International)

The top photograph shows one of Braniff's 'Flying Colors' schemes on 727-227, N435BN (c/n 21042) in the background in 1975, with 727-227, N508BN (c/n 19993) in the foreground painted by Alexander Calder to represent the red, white and blue of the Nation's flag in motion as a tribute to the 200th birthday of the United States. The identification of the airline did not appear on the jet, only the signature of Calder. Since there can be only one 'Flying Colors of the United States', all other aircraft in Braniff's fleet carried the name, 'Flying Colors'. A DC-8 was similarly painted. (Braniff International)

The bottom photograph shows Braniff's penultimate scheme on 727-227, N463BN, (c/n 21490) seen on finals to Los Angeles (LAX) on 1 November 1985 (Hans-J Schröder)

CONTINENTAL AIRLINES
U.S.A.

Continental Airlines, which is owned by the Texas Air Corporation (TAC), is the fourth-largest US carrier. Contintental was formed on 15 July 1934, when it began services as Varney Speed Lines. In May 1937, Varney purchased the Denver – Pueblo route of Wyoming Air Service and moved its headquarters to Denver. Later the same year, the name Continental Airlines was adopted. Development continued for a number of years until the 1955 award of the Chicago-Los Angeles route (via Kansas City and Denver) marked the company's full transition to a mainline trunk carrier. In October 1981 Texas Air, parent company of Texas International Airlines, acquired a controlling interest in Continental and on 31 October 1982, Texas International merged with Continental. Continental operates an extensive network of scheduled passenger services, covering eighty-two US domestic and forty-two international destinations in Australia, New Zealand, the UK, France, Mexico, Canada, Japan, Micronesia and Tahiti. In 1989, it added Seattle-Tokyo and Guam-Brisbane-Cairns to its routes. A new colour scheme has recently been adopted with a very thin dark blue cheat line under the windows, and a dark blue fin with gold globe insignia on a white and grey fuselage. The previous scheme was more colourful and interesting with red and gold cheat lines, gold fin with red and gold badge. The rest of the fuselage was beige with black roof titles.

The photograph shows 727-227 Advanced, N563PE (c/n 21045) in Continental's old scheme, seen landing at Washington National Airport, DC on 28 October 1987. (Adrian Balch)

DELTA AIR LINES

U.S.A.

Delta Air Lines, the Atlanta-based US major airline, absorbed Western Airlines in April 1987 to form the third-largest US carrier. The airline operates from six main hubs at Atlanta, Cincinnati, Dallas – Fort Worth, Salt Lake City and Los Angeles, with services provided throughout the USA to around 160 locations. International flights serve points in Canada, Bermuda, Bahamas, Puerto Rico, the UK, France and Germany. Delta Air Lines was formed in 1924 as the world's first crop dusting company and commercial passenger services began in 1929. In 1953 the airline merged with Chicago and Southern Airlines (founded in 1933) and in 1972 it absorbed

Northeast Airlines (also founded in 1933). Delta's first 727s came from the Northeast fleet, but have since been sold. Today, Delta is the largest 727 operator with 130 in its fleet. Colour scheme comprises a dark blue traditional cheat line with red and white trim above it and Delta's long-standing red, white and blue swept triangle insignia on the nose and fin. Delta roof titles are in black.

The photograph shows 727-232, N418DA (c/n 21271) on finals to Phoenix, Az. on 6 September 1977. This aircraft has been in service since January 1977. (Werner Gysin-Aegerter)

EASTERN AIRLINES
U.S.A.

Eastern Airlines was formed in 1926 as Pitcairn Aviation, starting a mail service between New York and Atlanta in May 1928. In 1929, the company was acquired by North American Aviation, the name being changed to Eastern Air Transport. In 1938 North American relinquished its holding and the company adopted its present title after some reorganisation. After absorbing several airlines, Eastern took over Air Florida's transatlantic route from Miami to London on 15 July 1985. Eastern was the first airline to introduce the 727 – on 1 February 1964 on the Miami – Washington – Philadelphia route. Eastern operated a vast network of scheduled services to over 100 cities in the USA, Bermuda, Canada, the Bahamas, the Caribbean and South America. The company was the pioneer of the 'Air Shuttle'. The service between New York, Washington and Boston,

which offered a no-reservation, guaranteed seat operation, was inaugurated in April 1961. Eastern was absorbed by the vast Texas Air Corporation in 1986, although retained its separate identity. It was a sad day on 18 January 1991, when Eastern ceased operations after sixty-four years' service. The final colour scheme on Eastern's 727s was an overall natural metal finish with a two-tone blue cheat line which swept up the fin, known as the 'hockey-stick' scheme. Titles were blue outlined in white.

The photograph shows 727-25, N8106N (c/n 18257) in 'Air Shuttle' markings landing at Washington National, D.C. on 28 October 1987. This was only the sixth 727 built and served all its life with Eastern since delivery on 30 January 1964. (Adrian Balch)

EVERGREEN INTERNATIONAL AIRLINES U.S.A.

Evergreen International is an established US supplemental cargo carrier with worldwide operating authority. It has recently added a Hong Kong – New York (JFK) – Columbus service. Known until 1975 as Johnson Flying Service, Evergreen International was founded in 1924. Other points served include Seattle, Portland, San Francisco, Los Angeles, Chicago, Dallas/Fort Worth. Boston, New York, Atlanta and Vancouver. Colour scheme comprises two broad dark green cheat lines divided by a thin white line. The white fin has a diagonal triple dark green stripe design with the Evergreen logo. Titles are also dark green. The photograph shows all-cargo 727-78F, N728EV (c/n 18974) with freight door open at Langley AFB, Va. on 21 August 1986. (Stephen Wolf)

FEDERAL EXPRESS

U.S.A.

Federal Express, a 'hub-and-spoke' all-freight carrier, specialises in the door-to-door express delivery of packages and documents. Formed in 1971, operations began in 1973, using a fleet of Falcon 20 business jets. Following CAB deregulation of air cargo in November 1977, authority to operate larger aircraft was received and a fleet of Boeing 727 and DC-10 aircraft was assembled.

Based at Memphis International Airport, an extensive network of nightly jet cargo flights filter out over every major city in the USA and Canada. The company offers a two-day service to Europe, and a three-day service to Tokyo, Hong Kong and Singapore. Documents and packages not exceeding 69kg (150lb) in weight or 330cm (130in) in length and girth are carried. The colour scheme is a striking purple upper half of the fuselage and fin and white lower half, divided by a thin red and white cheat line, broken by the titles 'FEDERAL' in white and 'EXPRESS' in red.

The photograph shows 727-252F, N205FE (c/n 22927) at Eppley Field, Omaha, Nb. on 17 October 1989. Named 'Robert Christopher', this was one of the last 727s built and has served with the airline since delivery on 1 September 1983. (Adrian Balch)

HUGHES AIRWEST

UNITED STATES OF AMERICA

Based at San Francisco, this airline was formed on 17 April 1968 as Air West Inc., by the merger of three local service airlines; Bonanza Airlines, Pacific Air Lines and West Coast Airlines. Air West changed its title to Hughes Airwest in April 1970, when Howard Hughes, through the Hughes Air Corporation, acquired control of the airline. Hughes Airwest operated to over fifty destinations in twelve western and mid-western states, four in Mexico and Edmonton and Calgary in Canada. The first 727 was delivered on 19 August 1976, followed by a further eight. On 1 August 1980, Hughes Airwest was merged with Republic Airlines, who in turn were merged into Northwest Airlines on 1 October 1986. Colour scheme was a striking overall golden yellow with medium blue stylised titles under the windows and a triple diamond insignia on the fin.

The photograph shows 727-2M7, N723RW (c/n 21202) named 'Spirit of Hughes Flying Boat' taxying out in a wintery setting at Calgary on 16 January 1977. This aircraft was delivered new on 1 December 1976 and has been passed on to Republic, then Northwest Airlines throughout the mergers and is in service with the latter today. (John Kimberley)

KEY AIRLINES

UNITED STATES OF AMERICA

Key Airlines, a subsidiary of World Corporation, is a US charter carrier with worldwide operating authority. Key specialises in on-demand and contract flying operations. Principal passenger destinations include Las Vegas, Florida, Mexico, Canada, Europe and the Caribbean. Formed in 1962, the airline currently operates ten 727s. The airline's recent claim to fame came when it was announced that they had been contracted by the USAF to fly F-117 Stealth crews between Nellis AFB, Nevada and the secret and isolated Tonopah Field base in the north of Nevada.

Colour scheme features a light grey upper fuselage and fin with red cheatline trimmed with red and blue, which sweeps up the fin, separating the white lower half. The cabin roof title 'KEY' is in blue and 'AIR' in red-outlined gold letters.

The photograph shows 727-22, N27KA (c/n 18859) landing at Las Vegas, Nv. on 19 February 1987. This aircraft was bought from US Air in July 1983 and is currently in service. (Adrian Balch)

NORTHWEST AIRLINES UNITED STATES OF AMERICA

Northwest operates an extensive network of scheduled passenger and cargo services in the USA and Canada. Transpacific services are operated to Hawaii, Tokyo, Osaka, Seoul, Okinawa, Taipei, Manila, Hong Kong, Shanghai, Guam, Singapore, Saipan and Bangkok. Transatlantic destinations include London, Glasgow, Frankfurt, Paris and Amsterdam. The carrier also serves Cancun (Mexico), Montego Bay (Jamaica) and the Cayman Islands. The airline was founded in August 1926 as Northwest Airways, and adopted the title Northwest Orient Airlines in 1934. The company was reorganised in 1985 to form NWA Inc, a holding company for Northwest Airlines and several other subsidiaries. The carrier merged with Republic Airlines in 1986. As the launch customer for the 747-400, the airline introduced the type into service on its transpacific routes in June 1989. Northwest began 727 operations on 1 January 1965 on the shorter domestic routes. Today, the airline operates nine -100 series and sixty-two -200 series 727s. Colour scheme was an all red fin, with black cheat line and cabin roof part white and part natural metal. The latest colour scheme has a red upper fuselage and fin, medium grey fuselage and white lower fuselage, divided by a thin black cheat line. Titles and fin logo appear in white.

The photograph shows 727-2M7, N729RW (c/n 21742) in Northwest's old colours seen taking off from Eppley Field, Omaha on 17 October 1989. (Adrian Balch)

PACIFIC SOUTHWEST   UNITED STATES OF AMERICA

Pacific Southwest Airlines of San Diego, California, was formed to operate scheduled services in California as the airline division of Friedkin Aeronautics Inc. of San Diego, which started in 1945. Pacific Southwest had a simple network, San Francisco – Los Angeles – San Diego. Airports served were San Francisco International, Oakland and nearby San Jose; Los Angeles International, Burbank, Long Beach and Ontario; and San Diego. Limited service to Stockton and also Fresno was also operated. 727s replaced Electras beginning on 9 April 1965, meeting competition from United 727s and Western 720s. The 727-200s displaced the earlier 727s and PSA was operating eight 727-200s when they were taken over by US Air in 1987. Colour scheme featured an orange and red cheat line and fin stripe.

Cabin roof initials 'PSA' appeared in black logo style and all PSA's aircraft wore a 'smile' underneath the nose.

The photograph shows 727-214, N556PS (c/n 21513) lifting off from Los Angeles (LAX) on 5 April 1980, still smiling. (Stephen Wolf)

PAN AMERICAN WORLD AIRWAYS UNITED STATES OF AMERICA

Pan American World Airways (Pan Am), ceased operations in 1991 after sixty-four years in service. Pan Am was formed in 1927 and by the early 1930s routes were operated to Central and South America, the Caribbean and Mexico. In 1932, the company formed an Alaskan division. In October 1936, trans-Pacific passenger flights had commenced, followed by flights to Europe in 1939, and to Africa in 1940. American Overseas Airways, a major Atlantic competitor, was taken over in 1950. In the ensuring decades, Pan Am had become one of the best-known names in aviation history. The airline played a vital role in commercial jet transport when, in 1955, it placed orders for twenty Boeing 707s and twenty-five DC-8s. The carrier was also the first airline to operate the Boeing 747, from 1970. Until it ceased operations, Pan Am was operating a route network of over 403,000km (250,000 miles), linking over 100 points in fifty nations. Although its headquarters was in New York, it based its

aircraft throughout the world. Following the demonstration of the Boeing-owned 727 at Tempelhof Airport, Berlin on 2 December 1964, Pan Am ordered eleven 727s – eight for the intra-German routes, and three for operation from Miami to the Caribbean. The 727s entered service on 1 April 1966, replacing eighty-seven passenger piston-engined DC-6Bs. In March 1975, Pan Am suspended most of its flights out of New York to the Caribbean and northeast and sold some of its 727-100s, but had built up a fleet of eighty-nine -200s series by the time of the company's demise in 1991. Final colour scheme was plain overall white with huge blue 'PAN AM' lettering along the fuselage, through the windows. The long-standing Pan Am globe filled the fin as big as they could paint it.

The photograph shows 727-227, N567PE (c/n 21243) lifting off from Boston on the evening of 29 August 1990. (Stephen Wolf)

PEOPLE EXPRESS   UNITED STATES OF AMERICA

People Express Airlines was formed early in 1980 by Donald C. Burr, former president of Texas International Airlines, together with four former TIA executives. The airline flew high-frequency, low-fare scheduled passenger services to several major cities in the eastern States, including Buffalo, Columbus, Boston, New York, Norfolk, Washington DC/Baltimore, Syracuse, Sarasota, Jacksonville and West Palm Beach. Operating out of Newark, New Jersey, transatlantic flights were operated by Boeing 747s to London-Gatwick. People Express was taken over by Continental Airlines on 1 February 1987. The colour scheme was based on an attractive overall beige fuselage with chocolate and deep red cheat lines running down the fuselage, upon which 'PEOPLExpress' roof titles appeared in beige. The beige fin was broken by the airline's logo of the outline of two peoples' faces.

The photograph shows 727-243, N573PE (c/n 21266) seen taxying out at Norfolk, Va. on 4 August 1985. This aircraft was bought from Alitalia in June 1984 and was transferred to the Continental Airlines fleet when they merged with People Express on 1 February 1987. (Stephen Wolf)

PIEDMONT AIRLINES   UNITED STATES OF AMERICA

Piedmont Airlines was taken over by US Air in 1987 and was fully merged into the airline by 5 August 1989. Piedmont was established in 1940 as an aircraft sales and service operator, and began scheduled local services in 1948 as Piedmont Airlines. The carrier acquired Henson Airlines in 1983, and Jetstream International Airlines in 1984. Both continued to operate as Piedmont subsidiaries. Piedmont served 123 cities in twenty-nine states and the District of Columbia plus Ottawa, Montreal, London (Gatwick), and Nassau. In June 1966, Piedmont placed an order for 737s, and arranged to lease a 727 from January 1967 for fifteen months pending delivery of the 737s. At the time of the merger with US Air, Piedmont was operating 34 727-200s. Colour scheme comprised a traditional medium-blue cheat line with white cabin roof and broad cheat line below the blue. Piedmont's winged insignia appear in blue on the white fin, with 'Piedmont' lettering below it in red.

The photograph shows 727-281, N863N (c/n 20510) taking off from Norfolk, Va. on 28 April 1985. Named 'Pride of Dayton', this aircraft was bought from All Nippon Airways in December 1983. (Stephen Wolf)

TRANS WORLD AIRLINES   UNITED STATES OF AMERICA

Trans World Airlines (TWA) was formed on 1 October 1930, as Transcontinental and Western Air, through a merger of a part of Western Air Express (the other part became Western Airlines), TAT-Maddux and Pittsburgh Aviation Industries. The present name was adopted in 1950 and since then the airline has built up an extensive network of scheduled services in the USA, connecting over 100 destinations. Transatlantic services are operated to a large number of points in Europe and the Middle East, with main traffic hubs being at New York (JFK) and St. Louis. In 1990 TWA was in financial difficulties and sold some of its fleet. Transatlantic services ceased and other services cut back pending an uncertain future for the airline. TWA's 727s went into service on 1 June 1964, replacing Convair 880s and Boeing 707s temporarily assigned to medium range routes and, ageing Constellations and Super Constellations.

The first route for the 727s was New York (JFK) – Indianapolis, followed on 5 June 1964 by Boston–New York–Indianapolis. 727-200s replaced the early non-fan 707-131s and some DC-9s. TWA currently operate fifty-six 727-200s and seventeen 727-100s. The original colour scheme featured a red cheat line, which started at a point aft of the cockpit and gradually got thicker toward the rear fuselage. Red 'TWA' in the airline's gold world insignia appeared on the white roof and fin. This scheme was changed in the mid-1970s to a double broad red cheat line below the windows and 'Trans World' titles on the white roof. The fin is mostly red with 'TWA' initials in white block capitals. The photograph shows 727-31, N853TW (c/n 18572) on a pre-delivery test flight prior to be handed over to TWA on 21 May 1964. (Boeing)

US AIR

UNITED STATES OF AMERICA

US Air was established in 1939 as All-America Airways, adopting the name 'Allegheny Airlines' in 1953. Operations began in September of that year with a unique 'pick-up' mail service. The present name dates from October 1979. Lake Central Airlines was acquired in 1968, Mohawk Airlines in 1972 and PSA and Piedmont Aviation in 1987. US Air completed the integration of Piedmont Aviation on 5 August 1989, creating an airline with over 420 jet aircraft and more than 48,500 employees. Scheduled passenger services are provided to 130 destinations in the USA, Canada, the U.K., Germany, Bermuda and the Bahamas. A number of smaller carriers operate Allegheny Commuter flights; these connect with main line US Air services at major points. US Air's 727s were acquired when Piedmont Airlines were taken over in 1987, the airline currently operating twenty-nine of the type. Colour scheme is based on a natural metal fuselage with a broad cheat line running beneath the windows segmented in three colours of red. The three tones of red sweep up the fin, on which 'US Air' lettering is in white. Cabin roof titles are in white-outlined red and maroon.

The photograph shows 727-2B7, N772AL (c/n 22164) taking off from Norfolk, Va. on 4 August 1985. This aircraft was delivered new in April 1981 and was sold to Sterling Airways in October 1986. (Stephen Wolf)

WESTERN AIRLINES   UNITED STATES OF AMERICA

On 1 April 1987, Western Airlines became a fully-owned subsidiary of Delta Air Lines. The company can trace its history back to July 1925, when Western Air Express was founded. An inaugural flight over a Los Angeles–Salt Lake City route was made on 17 April 1926, using Douglas M-2s. Prior to the merger, Western operated a vast network of scheduled passenger services over the western states. Destinations included Fairbanks and Anchorage (Alaska), Vancouver and Edmonton (Canada), Honolulu (Hawaii) Acapulco, Mexico City, Washington DC and New York. The 727-200s were used on medium-range routes, releasing 720Bs for longer routes such as those to Hawaii and Anchorage. Western acquired their 727s during financial troubles and only obtained the aircraft when circumstances improved, the first being ordered on 23 January 1968. Colour scheme was a plain white fuselage and fin with a broad red cheat line, commencing with a huge red letter 'W' running down the windows. 'Western' appeared in black lettering up the fin, raked in line with the leading edge.

The photograph shows 727-247A, N2807W (c/n 20579) over Mt. Rainier, Washington State, during a pre-delivery test flight in May 1972. This aircraft is still in service with Delta Air Lines, following the merger on 1 April 1987. (Boeing)

WORLD AIRWAYS   UNITED STATES OF AMERICA

World Airways was formed in 1948 to operate charters. In 1950, the carrier was purchased by Edward Daly, who held eighty-one per cent of the shareholding up to his death in January 1984. In 1979, World began scheduled low-fare transcontinental services linking New York (Newark) and Baltimore/Washington DC with Los Angeles and Oakland on 12 April 1979. Scheduled services to Honolulu, London (Gatwick) and Frankfurt were added in 1981. World Airways closed down all of its scheduled services, which accounted for the bulk of operations, in September 1986, and now concentrates on worldwide passenger and cargo charters and on overhauls at its Oakland base. World ordered 727s on 13 July 1966

for use on Logair military charter services, in accordance with a Military Aircraft Command decision to require use of jets on these services, replacing DC-6s for the most part. World leased its 727s extensively, especially to Japan Air Lines and sold others to Air Mali and Varig. They owned a total of seven -100 series and all have now been sold. Colour scheme was based on a traditional red cheat line with red 'WORLD AIRWAYS' titles on the white roof. The fin sported a red flash and red globe insignia.

The photograph shows 727-173C, N692WA (c/n 19506) on a pre-delivery test flight over Washington State in August 1967. (Boeing)

UNITED AIRLINES
UNITED STATES OF AMERICA

United Airlines was formed on 27 March 1973 as the management company of four pioneer airlines: Boeing Air Transport, Pacific Air Transport, National Air Transport and Varney Airlines (founded in 1926). Capital Airlines was absorbed in 1961. In February 1986, the company completed the purchase of Pan Am's Pacific division, including spare parts, bases and over 2,500 former Pan Am employees. Today, United is one of the world's largest carriers with five major traffic hubs operating at Washington (Dulles), Chicago (O'Hare), Denver, San Francisco and Tokyo (Narita). An extensive system of scheduled passenger and cargo routes is operated, linking 156 cities in the USA, Canada and Mexico, with international services to Bangkok, Beijing, Hong Kong, Manila, Osaka, Seoul, Shangai, Singapore and Taipei in the Far East, Melbourne, and Sydney in Australia, Auckland (New Zealand) and Frankfurt and Paris in Europe. In 1991, United took over Pan Am's routes network in Germany and transatlantic schedules to London. United put the 727 in service on 6 February 1964 on one daily San Francisco–Denver round trip, initially in a ninety-passenger, one class configuration. They were only the second 727 operator, following Eastern by days. Services were expanded to other cities to expedite the phase out of the DC-6s, Convair 340s and Viscounts. Later, 727-200s were ordered in October 1966, enabling United to phase out Caravelles and contributing to the run down of the 720 fleet. Initial colour scheme comprised a medium blue cheat line down the windows, under which was a broad gold-trimmed white cheat line. Black raked 'UNITED' titles appeared on the white roof and fin, the latter being accompanied by United's red and blue flash. In the early 1970s, United modified their colour scheme to replace the gold-trimmed white band by a thicker red-trimmed white band and called their aircraft 'Friendship-727'. The traditional red, white and blue national colours were finally disposed of in 1975, when United changed its colour scheme to the present livery. This comprised a broad tri-coloured cheat line of orange, red and blue on a white fuselage. The tail logo is a stylised red and blue 'U' on the white fin. Roof titles remain in black.

The first photograph shows 727-22, N7006U (c/n 18298) in United's original colours on a pre-delivery test flight in November 1963 (Boeing). The second photograph shows 727-222, N7625U (c/n 19542) in its original colours over Washington State in June 1968. (United Airlines)

The third photograph illustrates United Airline's current livery on 727-222, N725U (c/n 21402), which is seen near Seattle, prior to delivery in January 1978. United currently operates 104 727-100s and 28 727-200s. (Courtesy United Airlines)